Peter Orlando Hutchinson

The ferns of Sidmouth

Illustrated with impressions from the ferns themselves

Peter Orlando Hutchinson

The ferns of Sidmouth
Illustrated with impressions from the ferns themselves

ISBN/EAN: 9783742836618

Manufactured in Europe, USA, Canada, Australia, Japa

Cover: Foto ©Klaus-Uwe Gerhardt /pixelio.de

Manufactured and distributed by brebook publishing software
(www.brebook.com)

Peter Orlando Hutchinson

The ferns of Sidmouth

THE

Ferns of Sidmouth.

ILLUSTRATED

WITH IMPRESSIONS FROM THE FERNS THEMSELVES.

BY

Peter Orlando Hutchinson,

AUTHOR OF

"THE GEOLOGY OF SIDMOUTH," "NEW GUIDE TO SIDMOUTH,"

"HISTORY OF THE RESTORATION OF SIDMOUTH PARISH CHURCH,"

ETC.

SIDMOUTH:

PRINTED, PUBLISHED AND SOLD BY J. HARVEY,

FORE STREET.

1862.

Introduction.

THERE is nothing particularly new in this book unless it may be in the Illustrations, which are printed from the fern leaves themselves in a way which I believe has not been hitherto made public. The process is simply this. Take a fern leaf—cover one side of it with printers' ink— transfer it to the lithographic stone—and have it printed on paper. This is the process; but still, it is scarcely so brief or so easy as that. One of the best ways to put the ink on the leaf, is first to distribute it with a dabber made of cotton-wool tied up in a piece of silk, over a sheet of paper, and as thick as a good coat of paint. Upon this lay the leaf; and having placed a spare sheet of paper upon this, press it down with the hand, or in any other effectual way. Then draw off the top sheet of paper, and secondly the leaf, carefully. As the green surface is smooth, sometimes this must be repeated before the ink will take to it. There is now too much ink on the leaf, therefore make an impression on a spare piece of paper to get rid of the superfluous portion. If you have no lithographic stone at hand (as I had not) lay the leaf on lithographic transfer paper, cover it with some clean stout paper, and firmly draw a smooth ruler over it. This requires great caution for fear of blurring. A copying press would be a good thing. The leaf must be drawn off with care. Then write the name under, or anything else, with a steel pen and lithographic ink, rubbed up like Indian ink in a saucer. If the impression is satisfactory, roll it on a ruler or pipe of paper, and send it by post to the printer. The Illustrations in this book were thus impressed in Sidmouth, and sent into Exeter to be printed. Nicety and a little practice will ensure success. The plan might be useful for various other things besides ferns. This book contains a list of all the ferns I have hitherto found within a distance of three miles of Sidmouth. If you can find others and add to the list, pray let me know. Go and look.

POLYPODIUM VULGARE.

COMMON POLYPODY.

THE name of this fern comes from the Greek words Πολυς, *polys*, many, and Πους, *pous*, Ποδος, *podos*, a foot, traditionally from its numerous roots, though this would apply to innumerable other plants. Some say its roots resemble the polypi of certain fish, and hence the name. It is generally well to look into the derivation of a classical word used in the sciences, although it must be said that the applicability of such words is not always very obvious. The specific or second word *vulgare* is the Latin for common. There is no English designation for this fern in use, unless the term Polypody be so considered, though it is no more English than Polypodium. The farm labourers and uneducated country people in this neighbourhood use the latter word, which is very good Latinised Greek, and which would sound highly classic in their mouths if they did not pronounce it *Polypojum*.

The fibres of the root descend from the rhizome, or horizontal creeping stem, which is commonly about as thick as the finger, woody, yellow, covered with brown scales; and from it ascend the stalks of the leaves. The stipes or stalk usually occupies half the whole length of the frond, and is smooth and yellow, or light green. The leafy part is lanceolate in form, scarcely contracting below, the lobes oblong, slightly serrated, though not always, and bluntish at the ends. The sori, or patches of seed, are naked and circular, the marks by which the genus Polypodium is known. Each sorus (singular of sori) is placed at the end of the vein which nourishes it. The venation, or disposition of the veins in the lobes, now considered by Botanists an essential point of observation, is shewn in the plate. There is also a sketch of a magnified spherical spore case emitting the seed, with its upper half lifted off by the jointed elastic ring.

USES.—This fern made into tea with hot water, by which is meant a hot decoction, has been recommended as a purifier of the blood. Once upon a time in my mad youth, at that crude season when

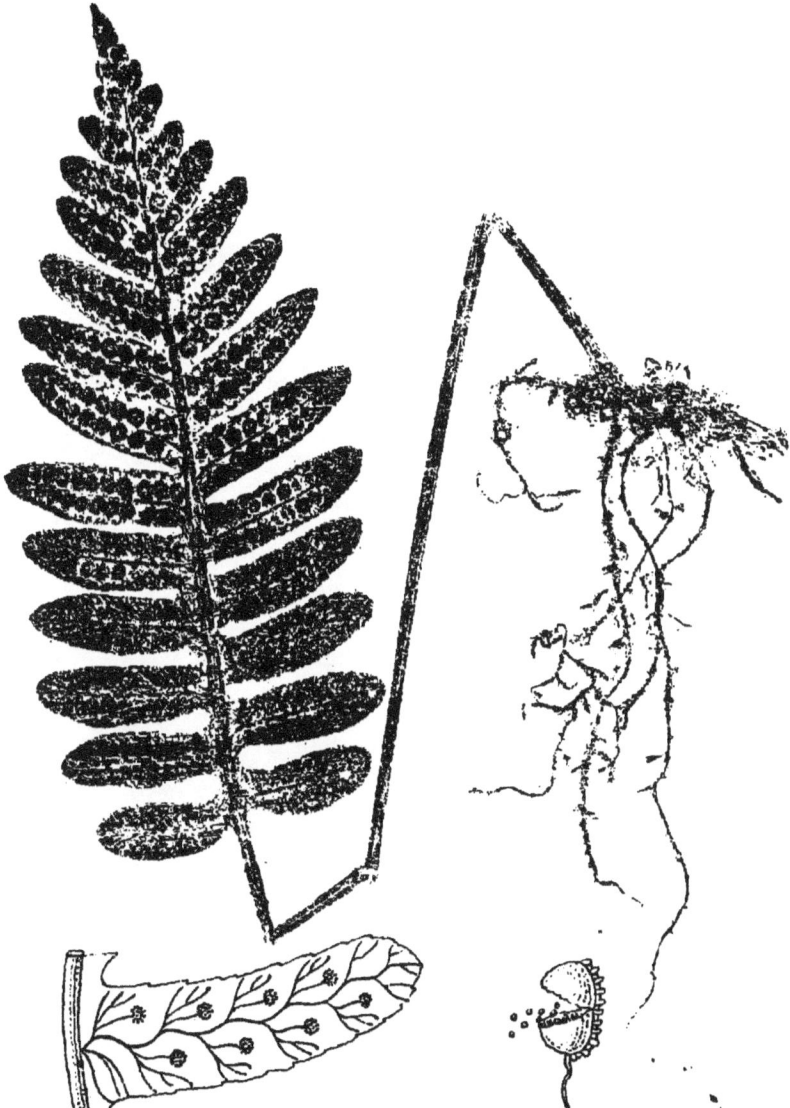

Veins in pinna.

Spore-case with elastic ring.

Polypodium vulgare.
Common Polypody.

people do many things for frolic or whim's sake, I undertook a course of this beverage. To the best of my recollection I drank one or two breakfast-cups full night and morning for several successive weeks or months, without milk or sugar. The taste, somewhat nauseous at first, soon ceased to offend. I forget whether I was ill when I began, or whether I was better when I left off. I cannot now say that it cured me, though I am certain it did not kill me. Woodville informs us that those plants are the best for medicinal purposes which grow on the stumps of the oak tree, though no good reason for this preference has been given. From the habit which this fern so frequently manifests of attaching itself to this tree, it has been styled the *Polypodium Quercinum*, from *Quercus*, an oak. As a demulcent and pectoral, its virtues have also been extolled. Joined with liquorice, its good effects have been experienced in coughs and asthmatic affections. It was once thought good in cases of mental aberration, especially in France; but considering that that indiscreet malady is still suspected to be lingering in some corners of that country, its complete efficacy is now questioned.

WHERE FOUND.—It is met with more or less plentifully in any of the lanes or hedges immediately out of the town. The nearest place is perhaps in the new Ottery road, or in the lane from Cotmaton to Jenny Pine's corner. If this does not do the seeker must go further.

CULTURE.—It thrives well in a garden or fernery. I have generally transplanted those specimens found growing in the actual ground, instead of those in the stumps of old trees, thinking that the roots would be less sensible of the removal. We must not shock the sensibilities of plants, any more than the susceptibilites of our friends, by transitions too glaring and abrupt.

POLYPODIUM VULGARE.

Varieties—LOBATUM, ACUTUM, CRENATUM, &c.

SEVERAL varieties of the Polypodium are to be found in the neighbourhood of Sidmouth, but it requires sharp eyes to detect them amongst a profusion of other vegetation, and in the secluded nooks in which they generally hide themselves. However, if they are worth finding they are worth seeking, and a little trouble spent in the hunt, will enhance the gratification of procuring them. The *Lobatum* is not uncommon. The fronds have two, three, four, or more fingers, and are fertile in seed-spots, even on the divided lobes. This fingered peculiarity should seem in a great degree to result from a luxuriance of growth, for it may be remarked that it occurs mostly in those plants which are the largest and most vigorous in appearance. I once gathered some in the high hedge on the left hand side of the Exeter road about a quarter of a mile beyond Stowford Gate, measuring nearly a yard from the root to the tip of the leaf. Specimens also used to grow in the left hand hedge 120 yards beyond the Broadway four-cross-way towards Bulverton. Also in most of the lanes on the east side of Bulverton Hill. Also two miles off towards Harcombe, from 100 to 200 paces in the left hand hedge beyond Snogbrook.

ACUTUM.—Nor is this variety uncommon. Indeed, it is easily detected amongst a wilderness of other plants in a hedge bank. The tapering points of the lobes of the frond are so obvious a feature as to render them discernible at the first glance: and there is a lightness and an elegance about the outline which is very pleasing to the eye. This variety is likewise fingered, and hence becomes, not only *acutum*, but moreover *lobatum*. In the hedges of most of the upper lanes on the east flank of Bulverton Hill, it will be seen in many places. Also 120 yards beyond Broadway, with the *lobatum*. Also near Bickwell and up Stintway.

CRENATUM.—The example in the plate is not so good as some which have been found near Sidmouth,

Acutum, Lobatum.
fertile.

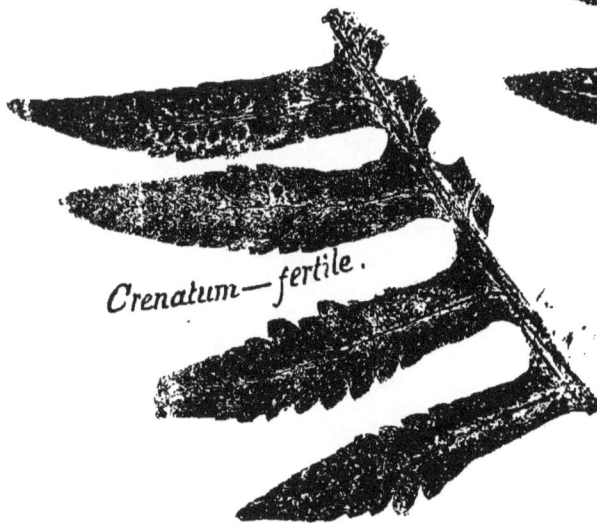

Crenatum—fertile.

Acutum.

Polypodium vulgare—*Varieties.*

Polypodium vulgare.
Variety Lobatum—*fertile.*

but it is the only one I had by me to print from. It is fertile. I cannot point to any exact spot where the enquirer would be sure to find it. Modifications of this variety seem to have been called *proliferum* and *Virginianum*. I have a frond from Buscombe Lane which combines the *acutum, lobatum,* and *crenatum,* all on the same stalk.

CAMBRICUM.—Mr. Edward Newman, in his *History of British Ferns,* p. 46, tells us that Dr. Greville found a gigantic specimen of this rare variety at Sidmouth. I have a frond from Boomer Lane which is rather irregular, but is more like a *cambricum* than anything else.

USES.—No difference between the varieties seems to have been made. The ancients thought that preparations of the Polypodium were serviceable if applied to limbs out of joint, and to chaps between the fingers and toes. The Hindoos boil a kind of Polypodium with their shrimp curries. A mucilaginous decoction of the plant was once in vogue for colds and whooping-cough. The fertile fronds are in some places gathered and dried, and boiled up with sugar when wanted. The root dried and powdered has been taken like snuff, under the idea that it would cure polypus. The powdered root has also been used to roll pills in. We are not informed whether such pills are more savoury.

POLYSTICHUM ACULEATUM.

[Acute Many-order, from the many lines of seed spots.]

COMMON PRICKLY SHIELD FERN.

SYNONYMS.—*Polypodium aculeatum, Aspidium aculeatum.*

THE student will experience some difficulty between this fern and the *P. angulare.* When their principal distinguishing characteristics are in each strongly marked it may be easy enough to see the differences when two such specimens are placed side by side. But when the guiding points of the *P. aculeatum* are a little relaxed, and when those of the *P. angulare* are not well defined, the two species assimilate to each other so much as to blend into a sort of indistinctness. When such negative, neutral, or mediastic plants as these are met with—and they are the most commonly met with—it becomes hard to assign them their true place and their true name. In the *aculeatum* the pinnules are narrower and decurrent, or merging into the rib of the pinna that supports them, but in the *angulare* they are attached by a short stalk: in the former they are of a darkish glaucous green, but are of a lighter colour in the latter: in the first, the pinnules form an acute angle at the point to which they are attached, whilst in the last, that angle is an obtuse one: and in the *aculeatum* the pinnules are somewhat convex on the upper surface, whereas, in the *angulare* they are flat. These seem to be small differences; nevertheless, if they are constant, they are enough to establish two acknowledged species. As long as a difference is constant it is sufficient, no matter how small it may be. As to the rest, the frond of the *P. aculeatum* is lance-shaped, or broadly lanceolate; texture harsh, but this can be only tested by comparison; the stipes or lower stem enveloped in rust-coloured scales; pinnules with an auricle or projection on the upper and outer side of each, serrate, spinulose; sori abundant, distinct, small, and brown; the indusium or cover is circular, which is characteristic of the genus *Polystichum,* and is fastened down by an attachment in the centre, as a button is sewn to a coat.

Pinnules of large specimen.

Pinna — ordinary size.

Seed spot, magnified.

Variety — lobatum, or lonchitidoides.

Polystichum aculeatum.
Common Prickly Shield Fern.

VARIETY—LOBATUM.—The distinguishing features in this variety are, that the frond is long and narrow, and that the pinnules are so much run together as to give the pinnæ a solid form. The pinnæ also, are in some cases placed so close together on the principal stalk as to touch or overlap; and this gives to the whole frond an entire appearance like the leaf of a tree. The example in the plate is by no means so long for its width as some I have met with, but it was the only one I had at hand to print from at the time. The more strongly this variety assumes the close or solid arrangement, whether in the structure of its pinnules, or in that of its pinnæ, the more it resembles the *Polystichum Lonchitis;* and hence it has, in this state, been called the *P. lonchitidoides,* or *lonchitis-like.* In obedience to the custom of the majority of Botanists, I have assigned this to the *P. aculeatum,* but Mr. G. W. Francis inclines to make it a variety of the *P. angulare,* alleging as a reason, that when the pinnæ break into lobes, as the larger and lower generally do, the lobes are invariably and very distinctly petioled or stalked. My own observation rather disposes me to his view. However, when we know, as stated above, that there is frequently great difficulty in deciding to which of the two species a certain fern may belong, from their great resemblance to each other, it is not surprising that persons of known experience should be perplexed in assigning the true place for a variety. The question, therefore, had better still remain open.

USES.—

WHERE FOUND.—In its most marked form it is not common. A root or two will be seen here and there ; but it abounds most in Cox's lane. Where is that? say you. Go half-a-mile out to the Turnpike on the Exeter road. Turn into the lane on the left immediately beyond it. That is Cox or Cox's. The fern is plenty from 50 to 100 yards up, on the right hand side—the variety on the left.

CULTIVATION.—It only needs shade and a dampish situation. It is nearly evergreen. The variety as well.

POLYSTICHUM ANGULARE.

[Angular Many-order.]

SOFT PRICKLY SHIELD FERN.

Syn.—*Polystichum setiferum, Aspidium angulare, Polypodium angulare.*

I have already observed how much this and the preceding fern are alike. Descriptions of minute shades of difference must necessarily be indistinct. There is nothing like looking at the objects themselves, and examining their various cast or contour of features with the natural eye. Any one who would wish to study the distinguishing marks depicted in a leaf or a plant, should take it in his own hands and scrutinise it closely. This will impress its prominent points on the mind, so that the plant when seen afterwards amid a thousand others, will be immediately recognised. How do we know a friend when we meet him in a crowd? We merely recognise those features which we have many times dwelt on before, and which have become so impressed upon us, that we can see them with our mind's eye, either in absence or in the dark. When seen in the midst of a multitude of other faces we light on them in a moment. We again look on those handsome eyes or that pug nose of our dear friend—those black locks or those carrots—that fair skin or that mahogany hide, as the case may be, and with pleasure we detect the well-known owner at the first glance. Make acquaintance, therefore, with the features of these ferns. Go into the lanes and hedgerows, and study them in their own homes. In default of that opportunity at the present moment, I must ask you to turn to the accompanying portraits, impressed from the very ferns themselves. A glance at the pinnules at the top of each plate, noting well the great differences which there are in their contour and shape, will do more than any description in making their points of non-resemblance apparent. In the pinnules of a large specimen, belonging to the *P. aculeatum*, those pinnules are attached to the rachis by a long point; they are rhomboidal in form, the auricle small, the top

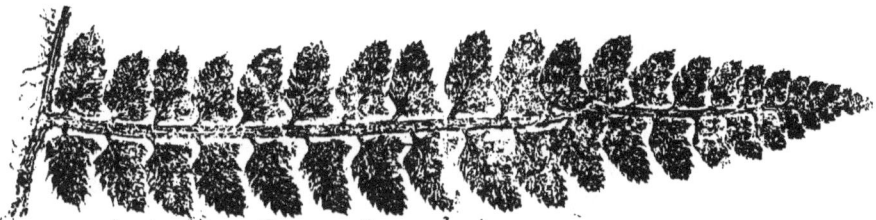

Pinna of a large plant.

Entire frond.

Variety Subtripinnatum

Variety Subtripinnatum
Larger specimen.

Polystichum angulare.
Soft prickly Shield Fern.

tapering : but in the pinnæ of a large plant in the plate referring to the *P. angulare,* the pinnules are broader, rounder, blunter at top, the auricles at their anterior bases larger, and they are all attached by short stalks. Though these descriptive differences seem to be trifling, nothing can be more dissimilar than the image which each one makes upon the eye. In all the noses which we see in human faces, how different they are, and yet they are all noses. And though no two of them may be exactly alike, how difficult it would be to describe by words the slight variations between them.

WHERE FOUND.—Where *not* found? It is abundant everywhere.

VARIETY—SUBTRIPINNATUM.—The little leaves or pinnules are themselves cut into little leaves, especially the lower ones, which are always the most highly developed. This multiplies the subdivisions of the whole frond once more; so that instead of being twice divided, it becomes three times divided, or *tri-pinnate.* The variety *decompositum* is very similar, if it is not still more divided. In other respects the *subtripinnatum* resembles the normal form, though its more lex nature causes it to droop with some grace.

USES.—

WHERE FOUND.—It is met with in localities favourable to vigorous developement. In Boomer Lane—the upper part of the lane which diverges opposite the Marino, and runs along the east flank of Peak Hill. The left-hand side going up Bulverton Hill Lane.

CULTIVATION.—When the common fern of this species is removed from the sterile hedgerow into garden ground, a more vigorous growth is induced, and thus it will frequently develope itself into the variety. This remark applies to most ferns. Cultivation turns many of them into their varieties.

LASTREA OREOPTERIS.

The word Lastrea was given in honour of M. de Lastre, a distinguished French Botanist.

THE HEATH SHIELD FERN, OR SWEET MOUNTAIN FERN.

SYN.—*Polypodium montanum, P. oreopteris, P. fragrans, Aspidium odoriferum, A. oreopteris, &c.*

THE fronds are vesica shaped, tapering upwards and downwards to a point each way, for the lower pinnæ diminish almost to nothing, and are set wide apart. They spring from the caudex or stump of the root in a circle, and dispose themselves like the feathers of a shuttlecock, generally attaining the height of from two to three feet. The stalk has a few scales on it below, but is hairy above, and is indented longitudinally by a channel on its upper side. The pinnæ are sometimes set opposite each other. They are pinnatifid, or cleft nearly to the rachis, to form the pinnules, which last are blunt, flat, and entire. The sori are arranged round the margin of each pinna, like a beautiful trimming of little brown beads. When they are in full show the effect is very striking. This fern has been mistaken for the *Lastrea filix-mas,* but it differs in being more delicate, smaller, and smoother: its pinnules differ also in not being crenate or notched, as in the Male fern. But the disposition of the sori settles every doubt in an instant. In the male fern they are ranged in a line at some little distance on each side of the midvein, and are confined to the lower half of the pinnule; whilst in the one under consideration they are smaller, set closer together, and are disposed round the edge even to the apex. And the fragrance is so strong and so grateful as at once to distinguish it from all others. A number of small, yellow, shining glands will be seen on examination on the under side. On being handled these give out a very agreeable balsamic odour. As to the venation, each lobe is traversed by a wavy mid-vein, which gives off alternate lateral ones to the sori at the margin.

USES.—

WHERE FOUND.—It is by no means common near

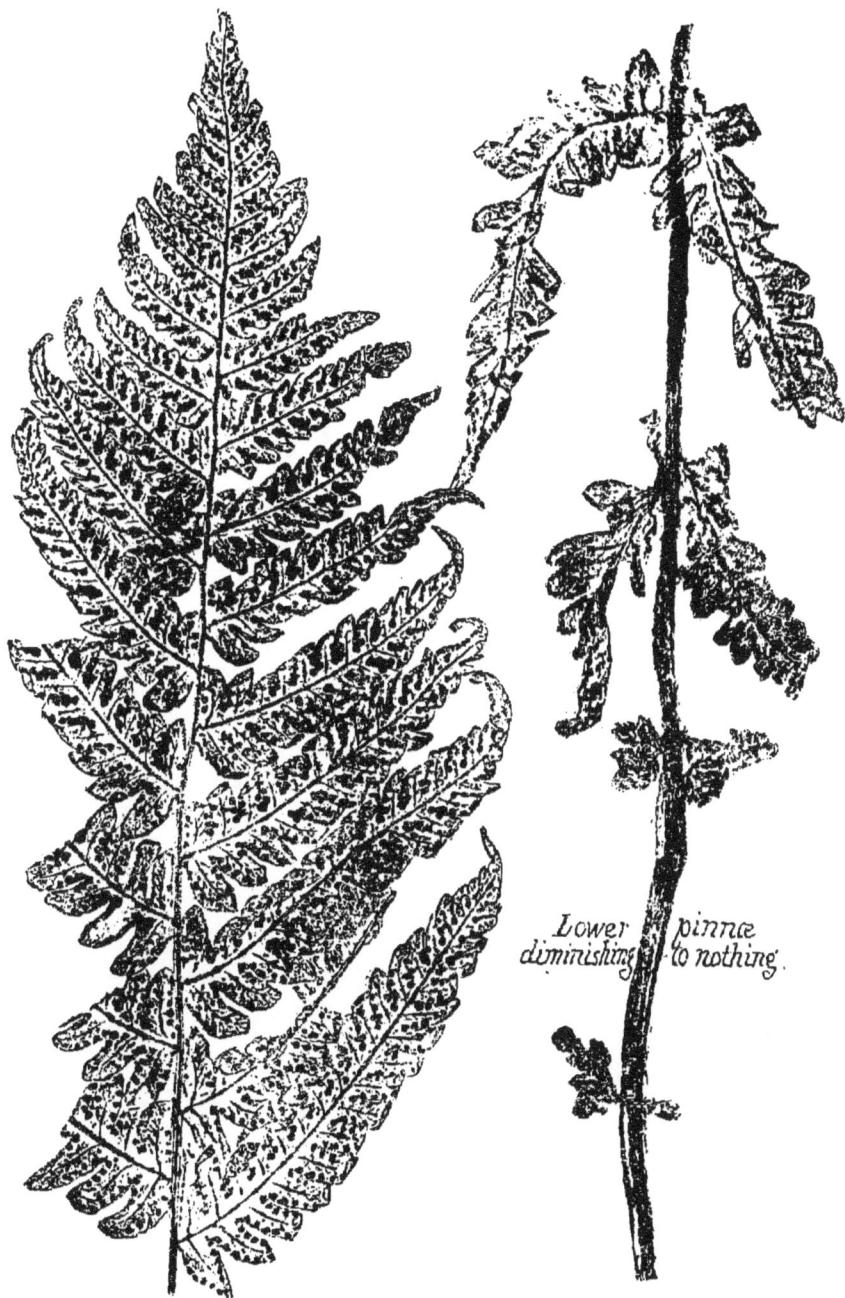

Lower pinnæ
diminishing to nothing.

Lastrea Oreopteris.
Mountain Buckler Fern.

Sidmouth. I hunted two years before I found it. This was well: for the satisfaction I at last derived in lighting upon my long-sought, was increased by the delay. Desires should not be gratified too soon, either in the seeking of ferns or fortunes. Is it not strange that delight can be found in a bog? Yet was it in a bog on Bucken Hill that a two years' search was rewarded by success. How shall I describe the spot? It is two miles and a half from Sidmouth. Go to Snogbrook, and then go up the lane to the saw-pit, the blacksmith's forge and the white cottages on the south flank of Bucken Hill. Mount the steep field between the saw-pit and the forge, and come out upon the open heath. At the foot of the heath is the swamp, and at the nearest or west end of the swamp there are a few roots in vigorous growth. Half a mile further on, at the eastern side of the hill, in Sweetcombe Valley, (pronounced Swetcum) at the top of the grass fields, are several scattered patches of the *L. oreopteris*. It seems to like this elevated situation.

CULTIVATION.—A loamy soil without any admixture suits it well. The warmth of a room or a greenhouse is too oppressive. The open air is more congenial. It must have constant moisture. The fronds appear in May, and do not survive the succeeding winter: indeed, they are generally cut down by the first sharp frost.

LASTREA FILIX-MAS.

[*Filix*, fern, *mas*, masculine.]

SYN.—*Polypodium filix-mas*, *Aspidium filix-mas*, *Polystichum filix-mas*, *Dryopteris filix-mas*, *Lophodium filix-mas*, *Filix non ramosa dentata.*

THE genus Lastrea is distinguished by having the indusium or cover of the seed spots of the shape of a sheep's kidney, attached to the leaf by the indented side. And this indented side is turned towards the mid-rib of the pinna. The root is large, tufted, and dark brown. Fronds broadly lanceolate, pinnate. Pinnæ alternate on the stalk, regularly tapering, deeply cleft. Segments oblong, blunt, crenate or notched, and close together. Sori confined to the lower halves of the segments, and to the upper half of the frond. Cover reniform or kidney-shaped, at first white and transparent, and afterwards brown. This is called the Male Fern merely from its robustness of make, as the *Athyrium filix-femina* is named the Lady Fern from its delicacy, and not from any anatomical structure such as might suggest such a distinction. The venation is easily studied in this fern. Each pinnule has an undulating mid-vein, which throws off alternate venules, simple or branched. Sometimes they are thrice-branched near the base, but simple near the apex. Where branched, the seed-spots are placed on the branch nearest the apex, but these seed-spots or sori, as remarked, are confined to the lower half or three-quarters of the pinnule.

USES.—An old belief ran to the effect that if a horse was ill, it would be cured if some of the root of this fern were put under its tongue. It was also thought that it had a great antipathy to the reed, and that the reed reciprocated this feeling—so that there was no love lost between them ; and that where one of these plants grew, the other would certainly not be found. The old physicians prescribed it internally in cases of intestinal worms. The young unexpanded fronds have been sold to the ignorant by designing persons, under the names of "Lucky Hands," and "St. John's Hands," as

Full size pinnules.

Variety — incisa, or erosa.

Lastrea Filix-mas.
Male fern.

preservatives against witchcraft. The plant was recommended to be burnt when gnats and other disagreeable insects were troublesome, as the smoke drove them away,—as if no other smoke would do. I have, however, seen the settlers in North America light fires of any wood that came first, before the doors of their log huts or other houses in summer, so that the smoke, playing about the entrance, prevented the mosquitoes from getting in. The male fern has been used in making beer; and as it contains both gallic acid and tannin, it would form a good substitute for hops. The inhabitants of Siberia sometimes boil it in their ale, in order to impart to it a certain flavour which they like. It is rich in alkali, and has been used in the manufacture of glass and soap, and in the tanning of leather.

VARIETY—*Incisa, erosa,* or *affinis.* This variety grows four or five feet high, and is large and handsome in appearance. Its pinnules are elongated to a point, or tapering towards the apex, are more deeply serrated, and are studded with an abundance of seedspots. *Incisa* means incised, or cut into, because the pinnules are serrated: *erosa,* eroded, corroded, or eaten into: *affinis,* akin to, resembling, allied to, that is, to the common sort.

WHERE FOUND.—It is sufficiently abundant in the neighbourhood, but is not confined to any particular locality, or hill, or dale, or lane. It is thinly distributed here and there all over this wide valley. To mention a spot or two—it will be found at 150 paces beyond Broadway towards Bulverton, in the left-hand hedge: about 200 paces beyond Stowford Gate, in the left-hand high bank. The variety occurs in Boomer Lane. It is not uncommon in the shady "goyles," as the deep Devonshire ditches are called, on the east flank of Bulverton Hill.

CULTURE.—Like a cosmopolite, it is at home anywhere. Only give it shade and a moderately damp situation. It is almost always cut down by the winter frosts, unless it is in a very sheltered situation. If it braves these, it will remain green till the young shoots appear in spring. The caudex or root endures to a great age.

LASTREA DILATATA.

[Lastrea dilated, spread out, broad.]

SYN.—*Aspidium cristatum*, [tufted, plumed,] *Lastrea multiflora*, [many-flower,] *Lophodium multiflorum*, [many-flower,] *Polystichum multiflorum*, *Polypodium dilatatum*, *Aspidium dilatatum*, and *Driopteris dilatata*.

THE frond of this fern is tripinnate, or thrice-cut, and not bipinnate, or twice-cut, as most of our ferns are. It is triangular, sometimes broad, drooping, dark green, and commonly about two feet high. Root tufted, black. Stem thick at the base, clothed with scales, thinner above, with a few scattered scales. The pinnæ opposite or nearly, pinnate, except the lower pair, which are doubly pinnate, toothed, and spinulose. The pinnules on each side of the lower pinnæ are unequal, the lower pinnules B. B. being much larger and more compound than the upper ones A. A. The sori abundant, the scales soon shrivelling up and disappearing from them. In disposition or arrangement they appear to range across the largest lobes, but lengthwise, on each side of the mid-vein on those which are less indented. The texture of the whole is fine and thin, and the edges of the pinnules curl over downwards. This fern soon withers after it is gathered. When specimens are intended for the Herbarium they should be put in the press as soon as possible. Amateur fern collectors are sometimes put to considerable shifts for presses, especially if the plants they may have gathered should happen to be of large size. In the domestic establishment, and amongst the ordinary furniture of a house, people must convert to their use whatever may best suit the purpose. The fronds should be placed between folds of blotting-paper, which paper should be changed and dried every day. A week ought to be enough to dry the ordinary kinds. If there is not a sufficient thickness of paper, the pinnules will have a cockled appearance, instead of being flat. As for a press, the paper and ferns may be laid on a level table, on which place a good stiff tray, and this can be weighted with books to any required

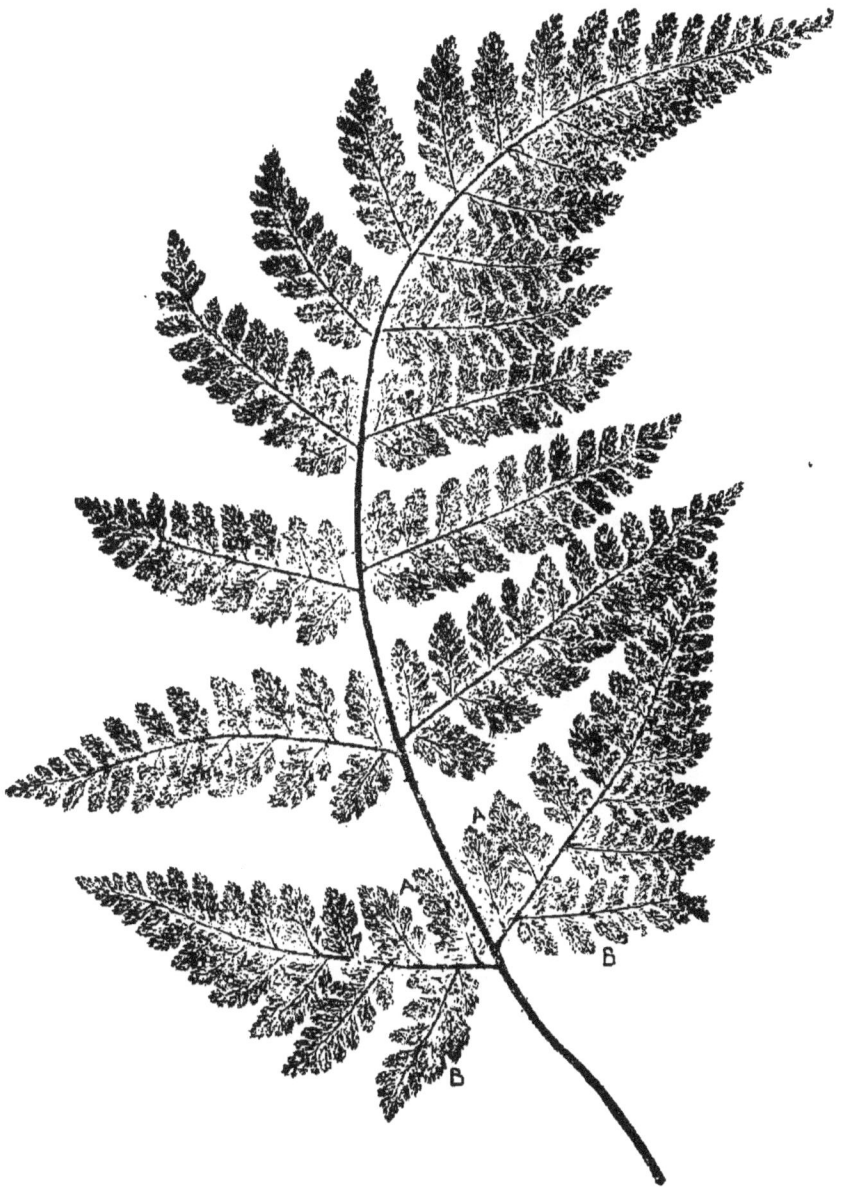

Liastrea dilatata.
Broad Buckler Fern.

amount. Some ladies place them between the cushions of the sofa, and then lie on them reading sentimental novels. Even the sternest fathers never find fault with their daughters for this arrangement: and if a gruff voice should ever cry out—"You lazy girl, why do you lie on the sofa all day long?" the natural answer will be—"I am only pressing my ferns, papa"—at which papa is quite satisfied. When dried they must be mounted. The best way is to employ two leaves of white paper, folio size or larger; and supposing these leaves open, fasten the fern on the right-hand side or page, and on the left-hand page write the name, date when procured, locality, description, or any other memoranda. Some fasten the fronds down with gummed slips of paper, some sew or tack the principal stems to the sheet with needle and thread, and some paste, or gum, or glue the whole under side of them, and then stick them down. Perhaps paste is the best adherent; it is more manageable than warm glue, and does not shine like gum. The best way to distribute the paste evenly over the back of the frond is to paste a sheet of spare paper, and lay the frond upon this pasted surface. On being lifted it will take up enough for the purpose desired. It must then be finally placed carefully upon its own sheet of paper, and dried under slight pressure. A portfolio of sufficient size is a good thing to keep the dried specimens in.

Uses.—

Where found.—Up the lane from Cotmaton towards Jenny Pine's Corner, opposite the stone wall of the Water-works of the Cotmaton Spring. Go from J. P's. Corner down the lane to Bickwell. It grows in the left-hand hedge by the gutter before reaching the cottages. In Bulverton Hill lane, (south-east corner of that hill) left-hand side it is abundant. By detached roots it grows all over the neighbourhood.

Culture.—In any ordinary soil it seems to succeed, if it has moisture.

ATHYRIUM FIFIX-FŒMINA.

[Opened female fern.]

THE LADY FERN.

SYN.—*Polypodium filix fœmina, Asplenium f. f., Aspidium f. f.*

THE name of this fern is a botanical compliment to the fair sex. As women are smaller of limb and fairer of face, and more delicate of structure than men, so this plant is more delicate in make among ferns than the rest of that tribe, and especially so than that one to which the name Male Fern has been given. The root is what is called tufted. The rachis smooth, green, or sometimes purple, and devoid of scales. Pinnæ or side branches, from twenty to forty pair, alternate, and tapering gradually to a point. Pinnules oblong, narrowing towards the point, serrated so deeply as almost to be cut in to the mid-vein,—which form is called pinnatifid. I am no advocate for employing long, learned, or scientific words if they can be done without: but in describing the parts of a plant, or the features of a plant, it is difficult to dispense with technical terms altogether. The word frond is preferred to the word leaf, although frond is only the Latin word for leaf. By the word leaf, as the leaf of an elm, we understand a solid flat expanse, whereas a fern leaf, except in the *Scolopendrium*, is cut up into many little leaves. Botanists, consequently, for distinction sake, prefer the name frond. Then, the centre stalk of the frond has side leaves, as in the *Polypodium*, called *pinnæ*, fins or wings. It would require many words to make a friend understand what part of the plant you meant, if you did not employ the word pinnæ; hence the convenience of adopting it. The same may be said of the expression *pinnules*, which are subdivisions of the *pinnæ*. And *pinnatifid*, a descriptive term used above, simply means, when a pinna (singular of pinnæ) is cut in almost to the mid-vein or stalk, but not quite. All arts and sciences have their technical terms, repulsive to new comers, who approach the entrance-gate of those studies: but after

Spore·cases
of Genus
Athyrium.

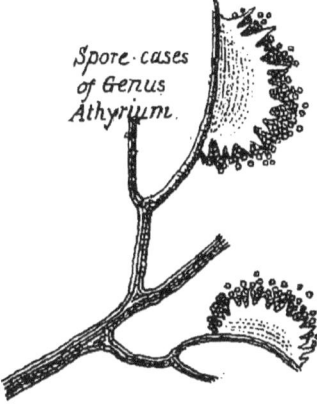

R.O. HUTCHINSON
FECIT

POLLARD, PRINTER, EXETER

Full size pinna.

Athyrium Filix·fœmina.
Lady Fern.

a little time they become more familiar, and then they are felt to be convenient. It is the over-use of these terms that will disgust the learner: the moderate use of them will be found almost necessary. Where was I? Oh, I was speaking of the Lady Fern. Well then, her caudex is tufted, and her rachis smooth, and her pinnæ pinnuled, and her pinnules pinnatifid. Her sori are numerous, covered with a scale or indusium of the form of a half-moon or segment of an orange, attached by its straight side to the side of the vein, the round outer part being fringed like eye-lashes. Advancing in growth, however, these coverings curl round till the two ends almost meet. In this state they become hippocrepiform. I apologise for using this lengthy word, which merely means horse-shoe-shaped; an expression which will answer every purpose.

As to Varieties, I cannot speak of them with certainty.

Uses.—From its softness, it is in some coast towns used to pack fish in.

Where found.—I have no-where seen it finer than in Bulverton Hill lane—that lane leading up to the south-east corner of the hill—and in the gutter on the left. Detached single roots occur here and there.

Cultivation.—It is easily cultivated. A lightish soil, with a sufficiency of shade and moisture will ensure success. The first autumn frost will cut down the fronds.

ASPLENIUM TRICHOMANES.

[Spleenwort Hairy, or excess of hair. Asplenium ασπληνον, *asplenon*, a medicine to cure diseases of the spleen, from *a* negative, and σπλην, *splen*. Trichomanes from τριξ, τριχος, *trix*, *trichos*, a hair, and μανος, *manos*, loose, long, from the long free hairs attached to the receptacles.]

COMMON MAIDENHAIR SPLEENWORT.

SYN.—*Asplenium melanocaulon, A. saxatile, Phillitis rotundifolia.*

THIS is a diminutive and pretty fern. It grows in tufts in the hedge banks, and rarely in old masonry near Sidmouth. The fronds vary in length from three to twelve inches. Stalk purplish black. Pinnæ ovate, green, slightly notched or crenate. Each one is traversed by a mid-vein, which gives off veinlets commonly divided in two, the anterior one of which bears the linear sorus. This sorus, linear at first, subsequently assumes a rounder appearance, when the abundant fructification causes the sori to become confluent, or to run one into the other, so that the reverse side of the pinnæ are almost covered. The countless multitudes of seeds that issue from the back of a single frond of a fertile fern is something beyond the capacity of the imagination to estimate. If they all grew the whole world would soon be clothed in a wilderness of ferns, to the exclusion of every other plant. Where, therefore, any vegetable (or animal) is the most prolific, the process of its destruction, by a very wise provision, is balanced with an equal pace, in order to keep it within due limits. Naturalists tell us that thousands of individuals have been counted in the roes of the herring, the cod, the salmon, and some others, and that the fronds of some ferns produce their spores by millions; and yet, as far as human observation goes, the sea is not fuller of these fish now than it was a hundred years ago, nor the earth fuller of ferns. It comes to this—that if the present enormous rate of production is only just enough to supply the constant demand, or the constant destruction, it is certain, that if fish were not produced by thousands

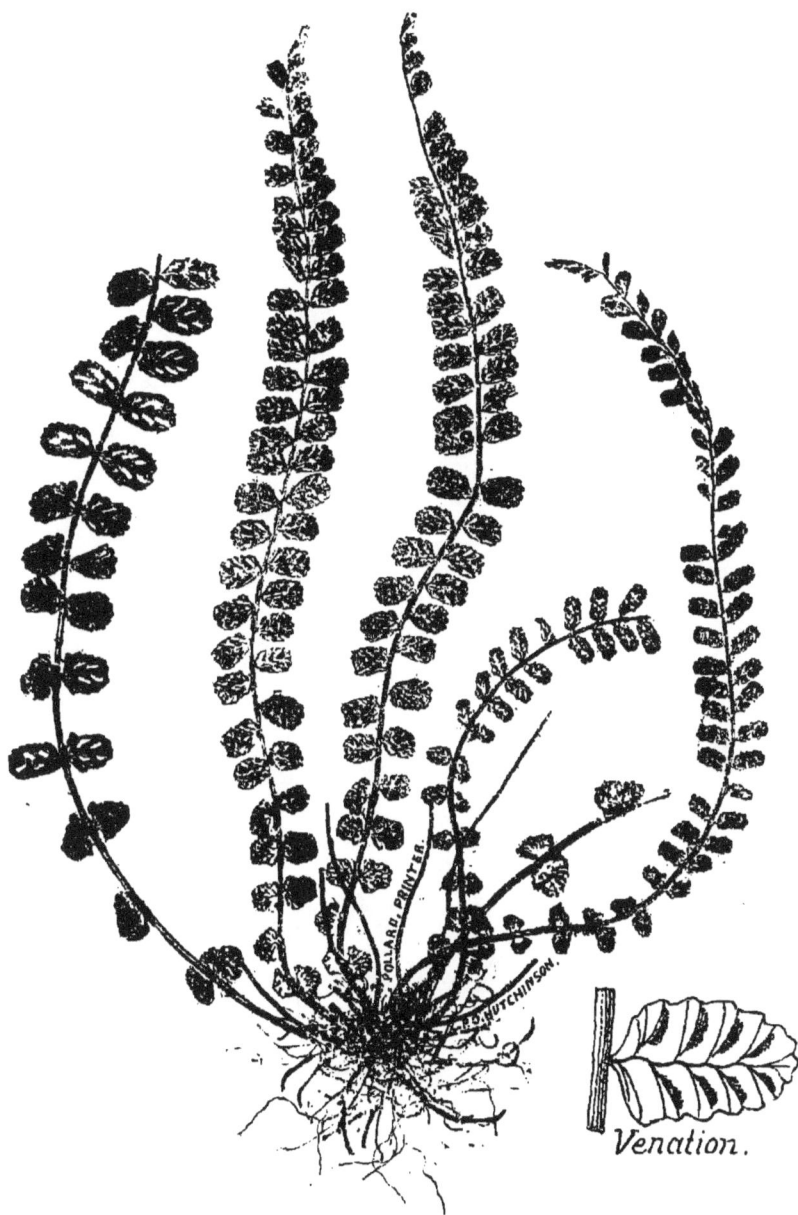

Venation.

Asplenium Trichomanes.
Common Maidenhair Spleenwort.

and ferns by millions, the world would soon be
without both fish and ferns. In the temperate zone
the proportion which ferns bear to other plants, is
one to seventy: in all England it is one to thirty-
five. England produces only one fortieth part of
the ferns found in the whole world. The number
of species in England is from forty to fifty, to which
may be added the numerous varieties.

Uses.—The Asplenium Trichomanes still holds a
reputation for medicinal virtues. This plant, together with four others,—the Hart's Tongue, the
Golden Maidenhair, the Wall-Rue, and Common
Spleenwort,—were termed the *Five Capillary Herbs*,
and were held in great estimation. The country
people near Sidmouth say, that tea made of it—by
which they mean a decoction—is good for coughs
and complaints of the chest. The old writers recommended it in obstructions of the bowels. As a
drink, a syrup made of the leaves with water forms
a pleasant draught.

Where found.—There is little or none nearer
to Sidmouth than a mile and a half, and only two
or three roots at that distance. At two miles off it
is abundant. Take a walk, for instance, out on the
Exeter road. Pass through Stowford Gate. Take
the first lane on the right, (Saltway Lane) about a
quarter of a mile beyond the Gate. There is plenty
in the left-hand hedge about 150 yards up. Also
in the lanes above, close under Beacon Hill. Also
in the long lane on the south side of Core Hill.

Culture.—Take it up with a ball of earth. As
it stands the winters, it has always a fresh and pretty
appearance.

ASPLENIUM MARINUM.

MARINE, OR SEA SPLEENWORT.

ALL writers seem to have agreed to call this plant by the same name, so we are not troubled with synonyms, as is the case with most others. The root is tufted, black, with stout fibres. Rachis, or stalk, black or brownish black, smooth, and shining, often bent at the base, free from pinnæ at the lower part, the first set opposite, the upper mostly alternate, slightly decurrent; upper side of each pinna auricled, or possessed of a projecting lobe; mid-vein commencing at the lower corner, and consequently not running through the middle. Pinnæ are firm in texture, toothed, and glossy green. The specimen from which the illustration was impressed, has the pinnæ rather longer and narrower than the normal shape. The sketch of the Venation in the plate shews that the oblong sori are produced from the anterior veinlet, as in the *Asplenium Trichomanes.* They range in two rows, one on each side of the mid-vein, and are set diagonally to it. The indusium, or scaly cover, opens along the anterior margin, or on that side nearest the point of the pinna, and furthest from the stalk. The plant is evergreen.

USES.—The thick consistence of the leaves will yield a mucilage, which was once thought to be a useful application to burns.

WHERE FOUND.—It grows in the yellow perpendicular rocks of the Dunscombe Cliffs. I see no reason why it should not grow in the yellow rocks at the upper part of Salcombe Hill, but it has not been detected there. In going to the Dunscombe Cliffs, surmount Salcombe Hill by the road; pass through the village of Salcombe, and proceed half-a-mile further to Dunscombe Farm. Go through the yard and straight out to the cliffs. A carriage can be taken this way. The farmer, however, discourages visitors through his premises. Or, there is another way—if on foot. Suppose yourself again

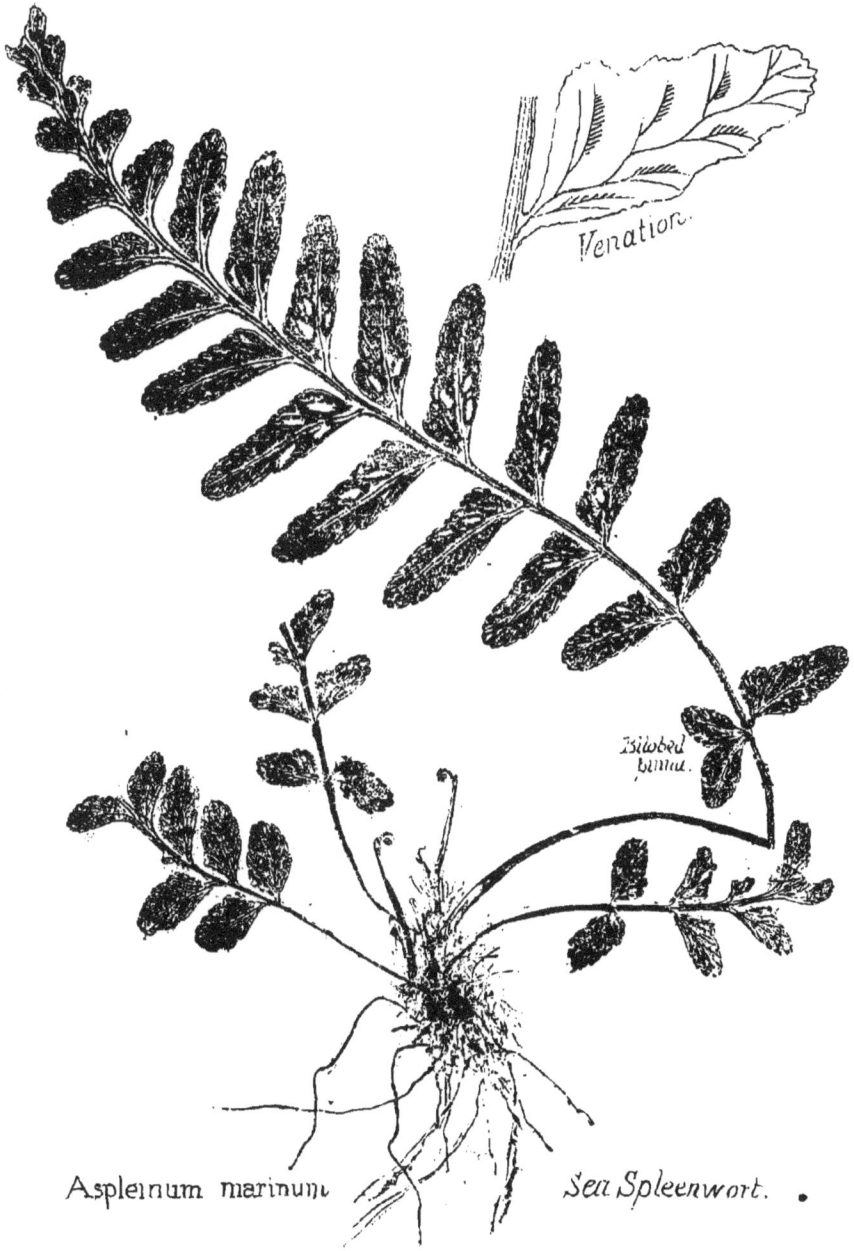

Venation.

3-lobed pinna.

Aspleinum marinum   Sea Spleenwort.

in Salcombe. A hundred paces or so after having passed the school a lane on the right is seen. Now remember—count five gates on the right from this lane, still going towards Dunscombe Farm, and the fifth gate you must stop at—considerably short of the Farm. Go over or through this gate, (they say you must never go over a gate when you can go through it) and crossing the grass field, you come out at the head of a lane which takes you down to the Sidmouth end of the Dunscombe Cliffs. At the end of this lane turn over the gate on the left: steer eastward all through this picturesque wilderness, famous for pic-nics, and descend by the edge of the cliff partly down towards the cottages at Weston-mouth, and turn in on the right to the steep cultivated patches of ground. You now see the yellow cliffs above your head. You will be fortunate if you are able to reach and extract a root or two of the fern. The specimen I give in the plate came from there. It grows at Budleigh-Salterton, and easy to get at, but this is beyond our limits. There is a red cliff running inland on the western side of the meadows of the estuary of the river. It grows in this low cliff near the ground. The best way is to go out to the Lime-kilns, and then turning your back on the sea, seek the spot.

CULTURE.—It should seem to love iodine or a saline atmosphere, for it does not always thrive if removed from the sea air. A peaty, sandy soil suits it best. Its fronds attain great length when under a glass, or nurtured in a greenhouse.

22

ASPLENIUM ADIANTUM NIGRUM.

[Ασπληνον ἀδιαντον *nigrum*, Asplenon adianton *nigrum*, —spleen-curer, water-refuser black. Adiantum was an herb, according to the old writers, that refused water, or on whose leaves, after a shower of rain, the water would not adhere. The second Greek word comes from α, negative, or not, and διαινω *diaino*, to wet: *quod folia ejus aquam respuant,* because its leaves refuse water: yet Brodœus says they will take it if it be gently poured on.]

BLACK MAIDENHAIR SPLENWORT, OR SHINING SPLEENWORT.

Syn.—*Tarachia Adiantum nigrum, Asplenium lucidum,*—bright, clear, or shining Asplenium. French name, *Doradille.*

THE length of the frond, including the stalk, according to circumstances, will vary from two or three inches (as on dry walls) to nearly two feet, as in damp, shady situations. It is triangular, ovate, or deltoid in form, and dark green and shiny in appearance. It is twice cleft: once on each side of the principal stalk to form the pinnæ; and secondly, the pinnæ are cleft in order to produce the pinnules. In some cases, however, the pinnules, particularly the lower ones, are again divided, thus making the whole frond thrice cleft or tripinnate. Rachis black, or dark brown below, and curved, the leafy part occupying the upper half. Pinnæ alternate. The spore cases on their first appearance have their proper linear shape, but after they have burst, and have become more fully developed, they frequently meet each other, so that the back of the leaf is almost entirely brown instead of green, thereby exhibiting a very marked appearance. Each pinnule has a mid-vein, from which proceed simple or branched veins, which bear the sori.

VARIEGATUM.—A variety of this name is now and then to be met with. Its peculiarity consists in colour. The pinnules of the fronds are edged and spotted with light yellow, giving the plant a very

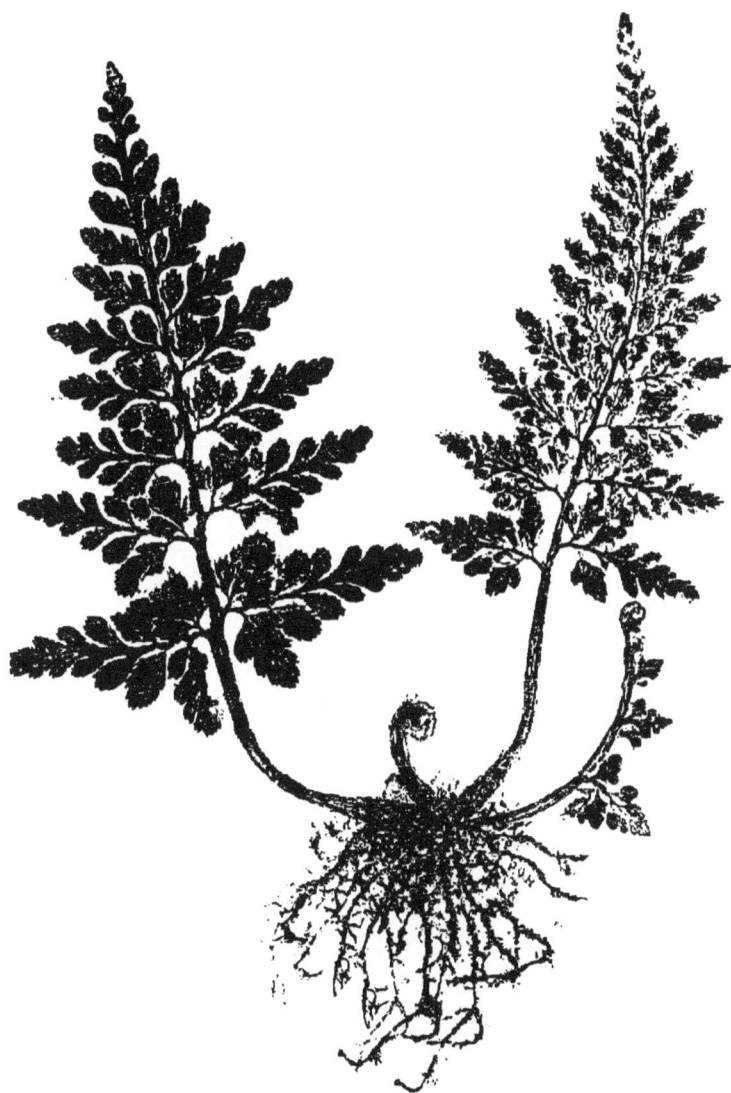

Asplenium Adiantum nigrum.
Black Maidenhair Spleenwort.

pretty effect. I have seen it in the left-hand hedge going up Salcombe Hill.

USES.—The old physicians (who had many primitive ways of attacking the ills that flesh is heir to) prescribed preparations of this fern in cases of cough, asthma, pleurisy, jaundice, obstructions of the spleen, and pains in the abdomen. It was likewise thought to be beneficial in scorbutic complaints.

WHERE FOUND.—First lane on right on the Exeter road, after passing the Toll-Gate, on the left side hedge going down to the river. Road up Salcombe Hill, after passing all the houses, in left-hand hedge. Going up Peak Hill, left-hand hedge, opposite garden wall of stone and red brick. New Ottery Road, quarter of a mile beyond Bulverton, in right-hand hedge. Moor, or Moor-Park Lane, near Mutter's-Moor, mostly south side.

CULTURE.—It is easily cultivated, and is a pretty plant either in a room or on rock-work. It thrives best in the open air. It prefers a sandy soil moderately moist. From being an evergreen it always looks neat and flourishing.

24

ASPLENIUM RUTA-MURARIA.

SPLEENWORT WALL RUE.

SYN.—*Scolopendrium Ruta-muraria, Phillitis R.-m., Amesium R.-m., &c.*

THIS is a small evergreen, not so abundant in this neighbourhood as it was formerly. Indeed, it has now become so scarce that I hardly know how to assist the collector in procuring specimens. The destruction of many old walls, which have been replaced by bran new ones, has worked terrible havoc amongst the favorite haunts of this tiny fern. The plant grows in tufts, the stems commonly being scarcely four or five inches high, the radicles of the root being as long. These wiry fibres insinuate themselves into the crevices of the walls so deeply, as to be exceedingly difficult to get out without injury. A splendid opportunity occurred in 1859. When Sidmouth Parish Church was pulled down and some of the memorials of the dead in the churchyard modified, I took down a course or two of the brick-work of Dr Blackhall's tomb, which was then a complete mine, and by thus separating the bricks extracted the whole mass of roots in one lump, and transferred these spoils to flower-pots. The specimen in the plate came from there. Let me remind fern-gatherers, however, that we don't pull down churches and tombs every day for the sake of giving them specimens. The stalk black near the root, but green upwards, occupies about half or more of the entire frond, the leafy portion being triangular in form. The pinnæ are alternate, oblong: the pinnules pear-shaped, heart-shaped, kite-shaped, broadly wedge-shaped, short, cleft at the sides, and toothed at the ends. A vein springs from the stalk, or point of attachment, and ramifies by subdivision towards the teeth. Two or more sori develope themselves near the centre of the leaf, and often appear to cover the under surface with fructification.

USES.—The genus *Asplenium* has been held in great regard as a remedy for diseases of the spleen— that organ of which nobody very clearly knows the

Asplenium Ruta-muraria.
Wall-rue Spleenwort.

use. The *Ruta-muraria* has also been used in affections of the lungs. It is the *Doradille des murs* of the French. Aσπληνον, *Asplenon, Asplenium ; lienis expers, curat lienosos,* it cures disorders of the spleen. Gerarde, in his Herbal of 1597, gives it another use. He says:—" The *Ruta-muraria,* or *Salvia vitæ,* is good for them that be troubled with paines or stitches in their sides."

WHERE FOUND.—At the present time it is not easy to say where. In by-gone years the village of Salcombe was full of it, but many old walls to which it clung have given place to new. Nevertheless, a careful search may still find it. I know two or three roots not a hundred miles from Sidmouth ; but if I tell you perhaps you will destroy them. But I know two or three others, and I will confide to you the secret. They grow in tufts from twenty to seventy feet high on the south side of Sidmouth church tower. You may have them and welcome ; but when you look up, perhaps you will say they are sour.

26

CETERACH OFFICINARUM.

[Ceterach of-the-shops. There is no English translated name for this fern, and it is not known with certainty from what country the name *Ceterach* originally came, nor the language from which it is derived. Some give it an Arabic derivation, and refer it to *Chetherack,* or to the words *shetr* or *chetr,* to cut, and *warak,* a leaf. Others give it a Keltic or Ancient British parentage, and quote the words *Cedor y wrach,* being the name of a plant spoken of by writers in the time of Charles I., meaning the Jointed or Double Rake, as if the leaf resembled two rakes placed back to back. Some suspect, however, that this name rather belonged to the *Equisetum fluviatile,* which others deny. Doctors disagree. *Cedor y wrack, Cedor wrack, Ceterach,* is a gradation of easy sequence, say the advocates for the ancient British derivation. We commonly call it the Scaly Spleenwort, or Scale Fern.]

Syn.—*Asplenium Ceterach, Grammitis C., Gymnogramma C., Notolepeum C., Scolopendrium C.*

Root tufted, fibrous. The fronds vary in length from three to six inches. The specimen from which I printed or impressed the example in the plate was the best I could procure at the time, though I was much dissatisfied with its small and stunted proportions. I could have procured a finer plant from a distance ; but in this book I admit nothing but what has been found near Sidmouth. The upper surface of the frond is of a deep green: the reverse is covered with brown chaffy scales, amongst which lie the sori. It is pinnatifid :—you know what that is, I suppose. I have explained once or twice, and you must assist me by remembering. The veins ramify from the base to the margin, where they cross each other. The sori lie on their anterior sides.

Uses.—Lamarck says it is operative and astringent, and has been used in spleen disease, as, indeed, all the Asplenons and their congeners have. Gerarde, the old writer on herbs in the time of Queen Elizabeth, observes, that any hardness or swelling of

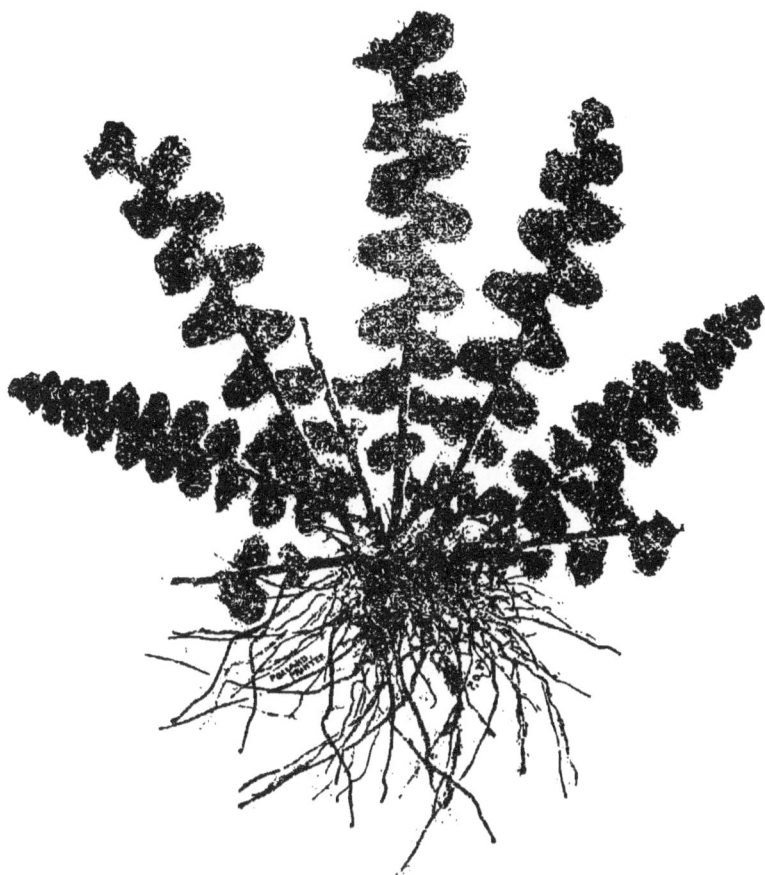

Ceterach officinarum.
The Scale Fern.

the spleen, or disease of the liver, could be cured with it.

WHERE FOUND.—And echo answers—where? One old wall is its only known habitat, and its admirers have well-nigh worked out its destruction.

> If you will tell me where to find the fern,
>   I'll get some specimens from that old wall:
> I'll take a knife and go, and ne'er return,
>   Till I have dug out mortar, roots, and all.
>
> No, will you now? Alas, my little fern,
>   Shall I teach others how they should annoy you?
> If that's the threat, perhaps they ought to learn
>   I cannot teach them how they should destroy you.

CULTURE.—It survives transplantation with difficulty. To give it the best chance, prepare some loamy earth mixed with brick rubbish, and plant it in a cold frame. It does not require much water.

SCOLOPENDRIUM VULGARE.

[Meaning Common Centipede. The first word comes from *Scolopendra,* a Centipede, the long brown seed patches, sloping away from the stalk, representing, to the eye of fancy, the legs of that insect. *Lingua cervina,* Hart's-tongue, is the Latin equivalent for our English name.]

ENGLISH NAME,—HART'S-TONGUE.

SYN. — *Asplenium Scolopendrium, Scolopendrium Phillitis,* &c.

ROOT tufted. Filaments long, strong, and black. Leaves numerous, from a foot to two feet high, strap-shaped, or ribbon-shaped; pointed above, heart-shaped below. Rachis, or stalk, scaly. Each elongated sorus, or seed patch, consists in reality of two sori placed face to face. When ripe they open in the middle throughout their whole length against each other, for the emission of the seed. The lower sorus is nourished by the upper fork of the vein below, and the upper sorus by the lower fork of the vein above. The fronds resist the winter frosts, and endure till the new ones unroll themselves in the spring. The example given in the plate is necessarily small, in order to accommodate it to the size of the book.

VARIETIES.—The most usual is that which is cleft at the end into two, three, four, or more lobes, expressed by the words *bifidum,* or *bifid, trifid, quadrifid,* &c., and many times cleft, *multifid.* This last presents the appearance of a bunch of frilling. Refer to the second plate. Keep your eyes about you, and you will find it near Sidmouth. Fern hunters must be wide awake. The variety *crispum* has the margin wavy like a frill got up on the Italian iron; but it loses this appearance by pressing. The variety *polyschides* is not common, so that increased vigilance is requisite if you are determined to find it—which I hope you are. Never give in. Go on, look again. If it is worth while to undertake the search, never give it

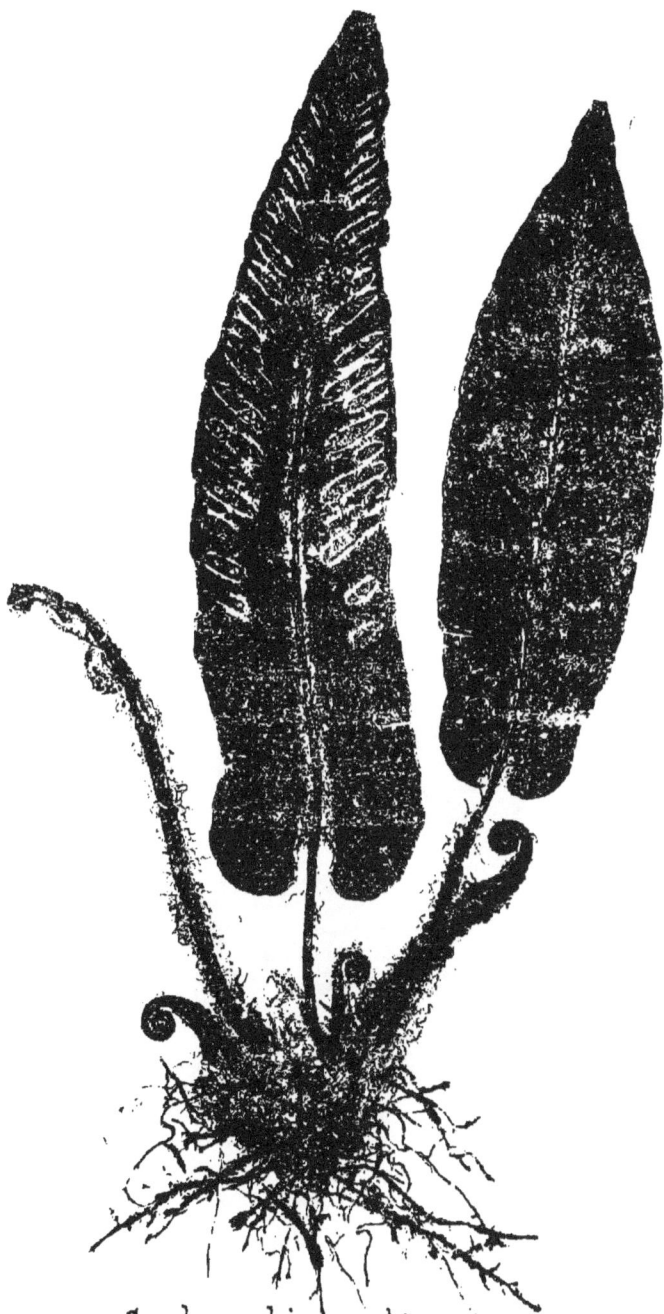

Scolopendrium vulgare.
Hart's Tongue.

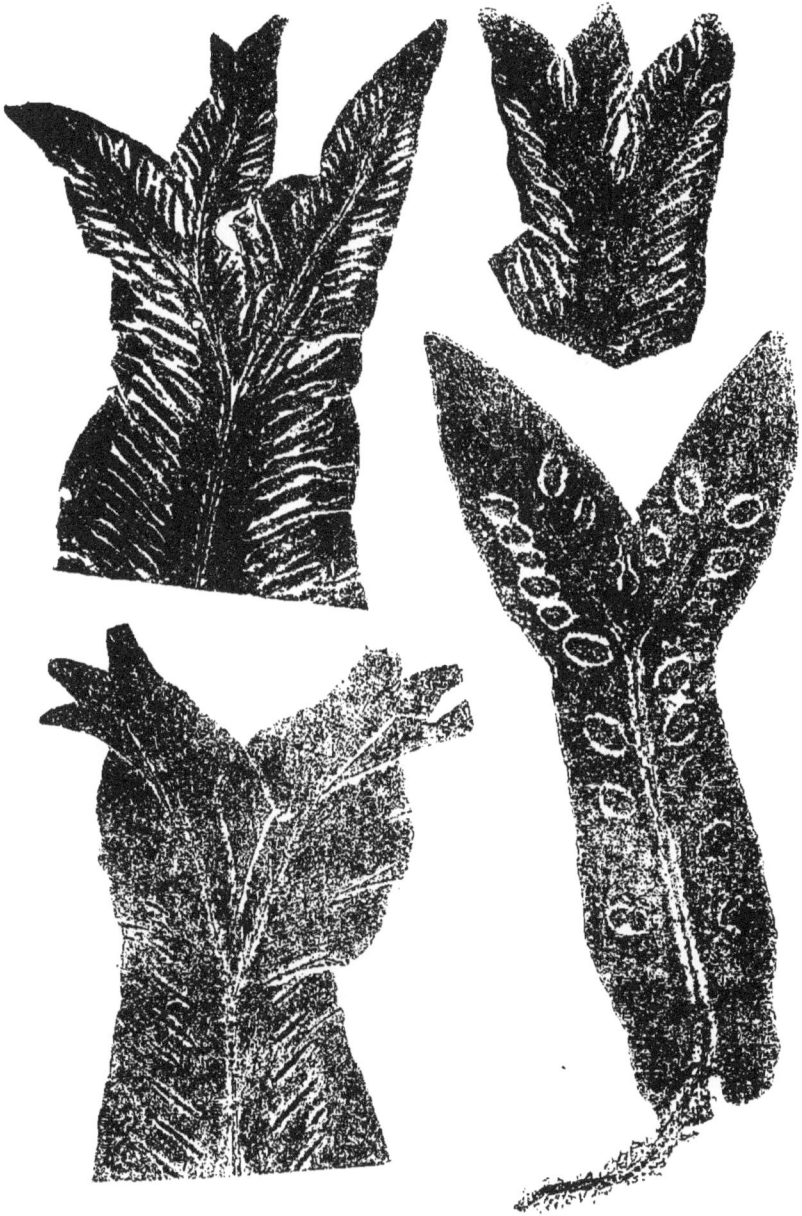

Scolopendrium vulgare.
Variety — multifidum, *forked, or fingered.*
Fertile fronds.

up till you have got what you want. In the *poly-schides* the frond is narrow, blunt at the end, and lobed, serrated, or crenate, along the margins. It is fertile.

Uses.—The genus is an astringent. The juice compounded with lard or other vehicle of that nature is made into an ointment, and used by the country people for wounds, scalds, or burns, and for assuaging hemorrhage. Taken internally the qualities of the plant act on the lower viscera. A decoction of the fern turns iron to a black colour. It has also been recommended as a strengthener of the bowels; as relieving the breast or windpipe; and as being favorable in removing obstructions of the liver and spleen.

Where found.—Where not found? The *Scolopendrium* is very abundant in this neighbourhood. Where every hedge-bank produces it, to mention localities is scarcely necessary. It is plenty in Cox's lane.

Culture.—It grows easily, especially in loamy soil and a shady spot. The varieties generally remain constant; and cultivation in a soil richer than what it has been accustomed to, will often convert the plain fern in its normal state into a variety, especially the *multifid*.

The best way to cultivate ferns in the house is to prepare some flower-pots by putting some potsherds in the lower part, and then filling up with light loamy soil to about an inch from the rim. The top surface should be rough and not pressed. On this sow the spores, and cover over with a piece of glass. The compost should be made of half peat earth and half loam, mixed with a quarter or an eighth part of clean sand or gravel. No manure is wanted. The water should be given at the bottom. The ripe ferns easily shed their spores in the room or greenhouse, and they soon grow if they alight on the sides of the flower-pots, or even on a wet piece of porous stone.

30

BLECHNUM SPICANT.

[Literally Spiked Fern : the word βληκνον, *bleknon*, signifying a fern.]

COMMON HARD FERN.   ROUGH SPLEENWORT.

SYN.—*Blechnum boreale*, [northern] *Osmunda spicant, Lomaria s., Asplenium s., Achrostichum s., Onoclea s., Struthiopteris s., Osmunda borealis.*

BY some old writers this fern has been confounded with the Pteris. But the marks of distinction that separate the two are these:—the Pteris bears its fruit close round the margin of the pinnæ, but the Blechnum parallel to the mid-rib, and near it: and whilst the frond of the Pteris is decompound, or much divided, the frond of the Blechnum is merely pinnate, or once cut, like the *Polypodium vulgare.* These distinctive marks are enough. There is but this one British species.

The root is tufted, and has stout, tough, black fibres. The barren fronds lanceolate, pinnæ oblong, rounded at the ends, lying close together. The fertile fronds grow taller, spring from the centre of the caudex, or stump of the root, are strap-shaped, tapering at each end. Their pinnæ curled in and contracted longitudinally, alternate, dilated at their base, separate from each other, under side covered with fruit. Sori continued in a line from the base to the point of each pinna, lying on each side of the mid-rib. The fertile fronds attain maturity in September, but die down on the approach of winter. The barren ones endure, and consequently they are evergreen. The venation is very intricate, but very beautiful on minute examination.

USES.—Ferns are of no use as food for man or beast. They contain so much of the bitter and astringent principle as to make them generally unacceptable. Even insects attack them but very little. A striking Geological argument has been used. When the coal beds are examined, it is found that they are chiefly made up of the decayed remains of an inconceivable amount of gigantic

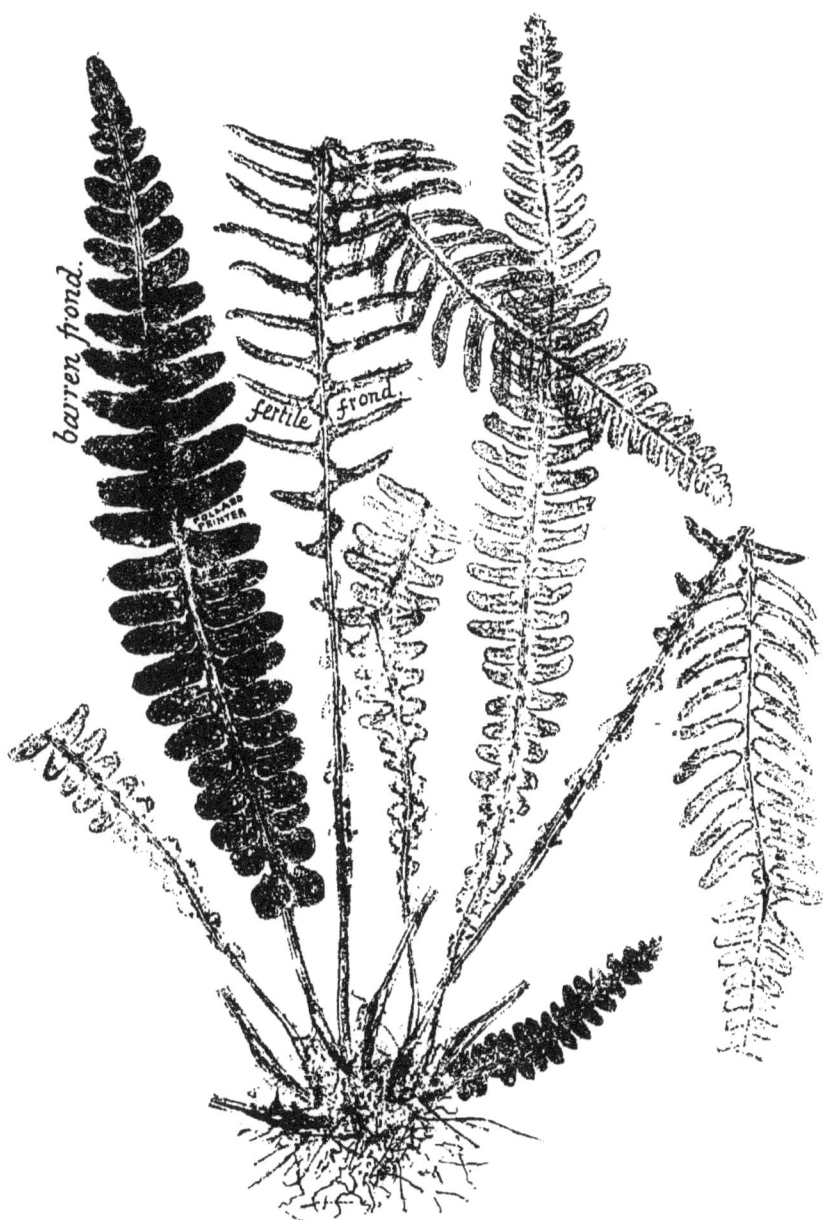

barren frond.

fertile frond.

Blechnum Spicant.
Common Hard Fern.

ferns. At that remote period the world seems to have been a wilderness of ferns. It mattered not that they were not good for food, for as yet no herbiverous animals had appeared upon the earth to eat them. Lying upon the coal measures are the Red Sandstone beds some 1,200 feet thick, all deposited by the slow process of subsidence: above that the Lias 70 feet: the Oolites 1,500: Wealden beds 900: the Green Sand 500: the Chalk 900; and then we come to the earth in which are found the first and earliest remains of the large herbiverous animals: wherefore it is said—As quadrupeds which live on vegetables had not yet appeared on the earth, it was not necessary that the vegetation of that period should have been calculated for their food.

WHERE FOUND.—Up Stintway, in the ditch on the left-hand side. Stintway is the green lane running from Lower Bickwell up the side of Peak Hill. On the east flank of Bulverton Hill, over the new Ottery road, at the bottom of the uncultivated part. In some other scattered localities by single plants, but in Harpford Wood, distant nearly three miles on the Exeter road, it is abundant, and easily found.

CULTURE.—The Blechnum prefers a damp situation, with plenty of shade, and will grow in peaty, gravelly, or clayey soil. A certain amount of peat, however, never seems to come amiss to the roots of ferns. Being a hardy plant, it succeeds better in the open air than in a close room or a greenhouse. As the barren fronds are evergreen they make a tidy appearance throughout the winter.

32

PTERIS AQUILINA.

[From πτερις, *pteris*, a Fern, from πτεριξ, πτερὸν, *pterix*, *pteron*, a feather. Newman proposed to name it *Eupteris*, emphatically *the* Pteris, to distinguish it from its allies. Aquilina from *aquila* an eagle, a postfix given by Linnæus, because he fancied he saw an eagle displayed in the section of the stem.]

COMMON BRAKES, OR BRACKEN.

RHIZOME, or root, about the thickness of the finger, black in colour, creeping underground horizontally. Rachis, or stem, thick and dark brown near the ground, green and tapering upwards, smooth and shining. Fronds from two to ten feet high, according to situation; nearly bipinnate when dwarf, but decompound when luxuriant. Lower pinnules pinnate and pinnato-pinnatifid, or twice pinnate. Sori, or seed spores, follow the sinuosities of the pinnæ, and lie all along under the reflexed edge. They are in fact placed under the hem. As there is but one British species of this plant its identification becomes easy. Some say this is the Θηλυπτερις, *thelypteris*, or Female Fern of Dioscorides and Theophrastus; and that if any of the female sex took it internally, or even walked over it, very injurious effects ensued. And now for an absurd anecdote, which you had better not read. Once upon a time when I was taking a country walk with some young friends, boys and girls of an advanced growth together, I saw two of the young ladies pulling up the stems of the Pteris aquilina, and cut them across through the black part with a pen-knife. Walking apart, they put their heads together, as if examining the stem so cut. If the result was not satisfactory another slice was cut off. I learnt afterwards that they were looking for the initial letters of their lovers' names in the section. Thus instructed, I have frequently since that time made the same experiment when I have been up among the brakes on the tops of these encircling hills. But in spite of many trials, I never could detect so much as the bare feature of a letter or an ampassy. Nothing, in

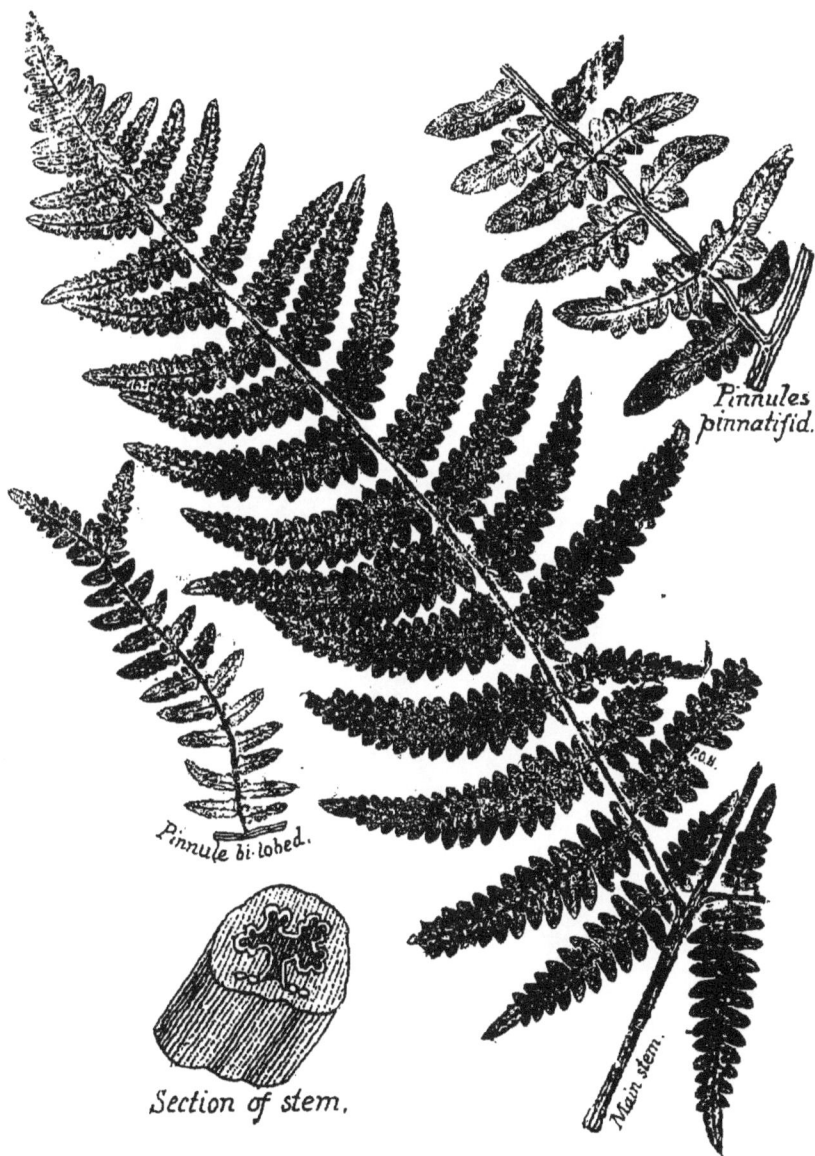

Pinnules
pinnatifid.

Pinnule bi-lobed.

Section of stem.

Main stem.

Pteris aquilina.
Common Brakes, or Bracken.

short, but a tolerably fair portrait of an oak tree. That, I think, is manifest enough. Indeed, some people call it "King Charles in the Oak."

Uses.—It is extremely rich in tannin, and has been employed for tanning leather. There is a moth called the Fern Moth, which feeds upon it. The ashes are abundant in alkali. To collect these, mix them with water, mould them into balls, heat them, and use the lye for washing linen, was long a practice in Wales. They are now better off for soap. The root contains much starch. Medicinally, the plant is taken in cases of intestinal worms. Decoctions of the rhizome and frond have been given in obstructions of the viscera and spleen. To lay a child with the rickets in a bed of the green leaves was looked upon as the sovereignest remedy. From its great yield of alkali, it has been employed in the manufacture of soap and glass. In Scotland it has been cut and thrown into the trenches when planting potatoes; and is said to produce good crops. In the Forest of Dean the young fronds boiled are used to feed pigs; though some say it is poisonous to cattle, and gives the trembles to sheep. In Silicia they employ the rhizome in the proportion of one third to two thirds of malt, in brewing their ale. In England the principal use of this valuable plant seems to be set forth in that absurd anecdote above, which I have requested you not to read.

Where found.—On all the hills encircling Sidmouth, and here and there in the lower grounds.

Culture.—The rhizome, or root-stock, creeps horizontally beneath the soil, sending its radicles downwards and its fronds upwards. The root, however, sometimes descends very deep. It has been found fifteen feet below the surface of the ground. The best mode of transplantation is to take up the plant in balls of earth. Like the Hart's-tongue, it dislikes a chalky soil. The young shoots spring up in May, but a frosty night will sometimes cut them off. The mature fronds soon perish before the approaching cold of the autumn.

34

OSMUNDA REGALIS.

[Which means Osmund Royal, or Regal.]

THE FLOWERING FERN AND ROYAL MOONWORT.

ROOT a stump, or tuber, from the top of which arise the fronds, which grow from two to ten feet high, green, twice pinnate. Rachis erect, smooth. Pinnæ distant, nearly opposite. Pinnules oblong, slightly crenate, or notched, and somewhat auricled. The leaves at the top of the frond produce their fructification from their substance, or become changed into seed; a fact so remarkable, so unusual, and so striking, as calculated to arrest the attention of the most careless observer. Pray consider the examples in the plate, which were impressed and printed from a plant gathered near Sidmouth. The left-hand example shows a branch which presents the common leafy appearance, but the right-hand one exhibits a state partly leaf and partly fruit. Even the pinnules are divided between fruit and leaf. The lowest one reveals a few seeds at the lower edge of the leaf; and the edge looks as if it had been eaten away where the seeds are. The two immediately above are half seed and half leaf, as if a sort of process of conversion or transformation had been at work. All the stalks which might have been the mid-veins of pinnules or small leaves, are stalks bearing bunches of spore coses. This is very remarkable and highly interesting, and would strike any one with wonder, if Nature were not full of wonders wherever we examine her works. And observe, that the spore cases sketched on an enlarged scale at the top left-hand corner, are without the elastic ring, which is common to all the ferns we have previously been considering. In the plate representing the *Polypodium vulgare* there is a sketch of a spore case with the jointed, elastic ring about it, by the elasticity of which ring the cover is lifted off. In the *Osmunda regalis* the spherical spore case is supported on a stalk, and it splits open into two halves working on a hinge. As to the derivation of the word Osmunda authors are not quite agreed. Some take it from the Saxon word *mund*, strength, and others from one Osmund, a

Spore-cases of Osmunda without the elastic ring.

Pinnules of.

barren pinna.

Leaves or pinnules entirely converted into seed.

Further formed

seeds forming on the edge of the leaf.

Osmunda Regalis —— Royal Fern.

waterman of Loch Tyne, who lived a thousand years ago. And in most of the works on Botany you will find that they tell a pretty tale, as how this Osmund hid his wife and daughter in a wilderness of this fern on an island in the lake to preserve them from the savage Danes, who were then over-running the land, and how the waterman's name Osmund was afterwards given to the fern. I hope it is true, because it is a charming story, but the whole affair is far too romantic to be admitted into a dull matter-of-fact little book like this.

USES.—In some counties an application of this fern, bruised, and soaked in water, is used in cases of sprains and contusions. Those who have been beaten or wounded sometimes drink a decoction of it. Withering says the root boiled is very slimy, and that in the north of Europe it is employed to stiffen linen.

WHERE FOUND.—It grows in several places among the hills above Sidbury and Harcombe. These places are hard to describe and hard to get at. I will, however, point out the easiest. Now attend. Go to Harcombe, two miles and a half off. When there, take the lane northward towards Sweetcombe, (pronounced *Swetcum*,) with Bucken Hill rising on your left. When at the distance of about half a mile from Harcombe, look out for an orchard on your left that comes right down to the lane. Then, at the top of the steep grass-field immediately beyond this orchard, you will find it growing in the wet ground close to the hedge. It also grows in the swamp out over the hedge.

CULTURE.—The roots should be transplanted by removing them in a ball of earth if possible. The newly-established plant ought to be placed near a pond or spring, as otherwise it will scarcely survive the change. It affects damp situations.

36

BOTRYCHIUM LUNARIA.

[From βότρυς, *botrys*, a bunch or cluster ; and *lunaria* from *luna*, the Moon, because the small leaves of the frond are lunate or half-moon shaped.]

Syn.—Linnæus called it *Osmunda lunaria*.

MOON-WORT.

This is another of the flowering ferns, though the expression is as incorrect, botanically speaking, as when applied to the Osmunda Regalis. It is a diminutive plant, attaining a height of from three to ten inches only. The root consists of a few coarse fibres, issuing nearly at right-angles from the central upright stalk. The stem is hollow at the base. It is an extraordinary and interesting fact, that if the stem be examined at its lower part, the rudiments of a small plant will be discovered, which are destined to spring up the succeeding year ; and still more extraordinary, that if a further and closer examination be made, the traces of a still more minute plant will be found, destined to flourish on the third year. The stem, in ascending, divides in two, the seed-bearing part being erect, the leafy portion slanting away from it. The leafy frond is divided into a number of fan-like pinnæ with uneven edges, the veins radiating from the point of attachment. The spike or spear is divided into branches, usually corresponding in number with the pinnæ of the frond, and these are again divided. This plant, so unlike a fern, is nevertheless classed with the ferns, for the same reason that the Osmunda Regalis is, namely, because it bears bunches of spore cases, and not true flowers, properly so called. These spore cases resemble those of other ferns, except that they are without the elastic rings, and, are borne on a stalk instead of on the back of the leaf. And whereas other ferns unroll themselves as they grow, this rather unfolds itself during the process of developement.

Uses.—The ancients believed that a piece of this plant put into the key-hole, would cause the lock to open : and that if horses grazed where it grew, their shoes would come off—such a strong effect had it upon iron. It is strange how such an absurd belief should have originated, and still more strange how it should for a day have maintained its ground, so easy would it have been to have disproved it. The Alchymists, we are told, thought they could turn quicksilver into pure silver with it. *Aside*—did they ever try? It was used as a remedy in dysentery. Made into ointment it was employed for cuts and wounds. Its qualities were better if gathered by moonlight.

WHERE FOUND.—Salcombe Hill. Said to grow on the sides of Bulverton Hill.

CULTURE.—Prepare a soil made of peat earth and sandy loam, which place in a cold frame. The roots should be transferred to it during the winter or early spring, before revivification has commenced. The soil should be kept only slightly moist. But some Botanists declare this plant to be a grass parasite; and that if its radicles, at the time of transplantation, be separated from the roots of the grass, among which they associate themselves in the meadows, it will not long survive the separation. Hence it has been recommended to take up a sod with the plant in it, which sod, grass and all, may be fitted into a pot or box, without manure, and placed in the open air. The grass must be kept short with scissors.

OPHIOGLOSSUM VULGATUM.

[From ὄφις, a serpent or adder, and γλῶσσα, *glossa*, a tongue. *Vulgatum*, common. Hence the English name Common Adder's-tongue. The farmers in this neighbourhood usually call it Adder's-spear, from the spear or spike which bears the seed.]

ADDER'S-TONGUE.   ADDER'S-SPEAR.

THIS little plant in many points resembles the Moonwort. From the perpendicular root spring the horizontal wiry radicles. The stem is succulent, and rises to the height of eight or ten inches. About half-way up or more, it divides into two fronds. The barren one is oval, and not unlike a leaf of the lily of the valley. It tends somewhat to a point at top, and is traversed by veins which ramify and frequently anastomoze. The fertile frond rises erect. Though this may look like a seed-bearing stalk, it has as much the nature of a frond, or series of fronds on a stalk, as what we see in the Osmunda Regalis. A reference to the plate shewing the Osmunda, exhibits stems bearing seed pods, and even leaves as if half converted into seed. Those which are all leaf may be termed barren fronds, whilst Botanists have designated the others fertile fronds. And this mode of expression has been applied to the *Ophioglossum Vulgatum*. The stalk of the spike does not bear a flower; it bears a cluster of seed in place of a leaf. At its upper part it is contracted into a sort of flat stem, on each side of which run a string of small beads, or branches bearing them, which are the spore cases. They are without the elastic ring. When ripe they split open for the emission of the seed.

The *Ophioglossum Lusitanicum*, or Dwarf Adder's Tongue, has never been found in England. It is common in the south of Europe, and has within the last few years, been met with in Guernsey. Botanists have been encouraged to hope that it might be found on the South Coast of England. During the early part of the year, I have made many diligent searches on the hills round Sidmouth, but hitherto without success.

Uses.—By the country people of Devonshire it is used in making Adder's-spear ointment, which is the sovereignest remedy for an inward bruise or an outward cut, allaying inflamation and healing all wounds. The component parts are Adder's Tongue, Boy's Love, House Leek, and Elder Blossom. These are boiled in fresh butter, strained, and potted for use. And it cured the bites of Adders, say the following lines—

> For them that are with newts or snakes or adders stung,
>
> He, seeking out an herb that's called Adder's Tongue,
>
> As nature it ordained its own like hurt to cure—

Where found.—Sides of Salcombe Hill. Not plenty, but said to grow in grass-fields on the slopes of several neighbouring hills.

Culture.—The management is the same as that described for the *Botrychium lunaria*.

So much for The Ferns of Sidmouth. I hope that your diligence and your vigilant eye may discover others, and add to the list above given. The use of a microscope is of course a great assistance as facilitating minute examination; but in default of this, a pocket case of three lenses is a very good substitute. Such a set, mounted in black horn with sides of mother-of-pearl, purchased in Regent Street, would cost twelve or thirteen shillings; whereas the same article obtained of Messrs. Chadburn, (brothers,) Opticians, Sheffield, may be had for three-and-sixpence. To this add the postage. Now go to work.

40

INDEX.

PAGE.

POLYPODIUM VULGARE. Common Polypody 2

———— Lobatum, Acutum, Crenatum, &c. .. 4

POLYSTICHUM ACULEATUM. Common Prickly Shield Fern. 6

———— ANGULARE. Soft Prickly Shield Fern. 8

LASTREA OREOPTERIS. The Heath Shield Fern, or Sweet Mountain Fern. 10

———— FILIX-MAS. 12

———— DILATATA. 14

ATHYRIUM FILIX-FŒMINA. The Lady Fern....... 16

ASPLENIUM TRICHOMANES. Common Maidenhair Spleenwort. 18

———— MARINUM. Marine, or Sea Spleenwort. 20

———— ADIANTUM NIGRUM. Black Maidenhair Spleenwort, or Shining Spleenwort. 22

———— RUTA-MURARIA. Spleenwort Wall Rue. 24

CETERACH OFFICINARUM. 26

SCOLOPENDRIUM VULGARE. Hart's Tongue. 28

BLECHNUM SPICANT. Common Hard Fern. Rough Spleenwort. 30

PTERIS AQUILINA. Common Brakes, or Bracken. .. 32

OSMUNDA REGALIS. The Flowering Fern and Royal Moonwort. 34

BOTRYCHIUM LUNARIA. Moon-wort. 36

OPHIOGLOSSUM VULGATUM. Adder's-Tongue. Adder's-Spear. 38

HARVEY, PRINTER, SIDMOUTH.